GARDENS ON MARS!

PROJECT

Continued from page

Continued to page

SIGNATURE

DATE

Witnessed & understood by

DATE

Top Secret Information

Continued from page

Continued to page

SIGNATURE

DATE

Witnessed & understood by

DATE

Top Secret Informatio

Continued from page

Continued to page

SIGNATURE

DATE

Witnessed & understood by

DATE

Top Secret Information

Continued from page

Continued to page

SIGNATURE

DATE

Witnessed & understood by

DATE

Top Secret Information

Student/Plant	Date	Soil pH Level	Comments

Notes

GARDENS ON MARS!

Continued to page

SIGNATURE

DATE

Witnessed & understood by

DATE

Top Secret Information

Notes

_____

_____

SIGNATURE _____   DATE _____

GARDENS ON MARS!

PROJECT

Continued from page

Continued to page

SIGNATURE		DATE
Witnessed & understood by	DATE	**Top Secret Information**

Continued from page

Continued to page

SIGNATURE   DATE

Witnessed & understood by   DATE

Top Secret Information

GARDENS ON MARS!

PROJECT

Continued from page

Continued to page

SIGNATURE	DATE

Witnessed & understood by	DATE	**Top Secret Information**

Continued from page

Continued to page

SIGNATURE

DATE

Witnessed & understood by

DATE

Top Secret Information

Notes

_____

_____

_____

SIGNATURE _____ **DATE** _____

GARDENS ON MARS!

PROJECT

Continued from page

Continued to page

SIGNATURE

DATE

Witnessed & understood by

DATE

Top Secret Information

GARDENS ON MARS!

PROJECT

Continued from page

Continued to page

SIGNATURE		DATE
Witnessed & understood by	DATE	**Top Secret Information**

Continued from page

Continued to page

SIGNATURE

DATE

Witnessed & understood by

DATE

Top Secret Information

GARDENS ON MARS!

PROJECT

Continued from page

Continued to page

SIGNATURE

DATE

Witnessed & understood by

DATE

Top Secret Information

Continued from page

Continued to page

SIGNATURE

DATE

Witnessed & understood by

DATE

Top Secret Information

GARDENS ON MARS!

Tables

Notes

_____

_____

SIGNATURE _____ DATE _____

Continued from page

Continued to page

SIGNATURE

DATE

Witnessed & understood by

DATE

Top Secret Information

Continued from page

Continued to page

SIGNATURE

DATE

Witnessed & understood by

DATE

Top Secret Information

Continued from page

Continued to page

SIGNATURE

DATE

Witnessed & understood by

DATE

Top Secret Information

GARDENS ON MARS!

Continued from page

Continued to page

SIGNATURE

DATE

Witnessed & understood by

DATE

Top Secret Information

GARDENS ON MARS!

Grid (x,y) Coordinates

Area:_____ Component: _____ Quantity: _____

(,) #___ (,)	(,) #___ (,)	(,) #___ (,)	(,) #___ (,)	(,) #___ (,)
(,) #___ (,)	(,) #___ (,)	(,) #___ (,)	(,) #___ (,)	(,) #___ (,)
(,) #___ (,)	(,) #___ (,)	(,) #___ (,)	(,) #___ (,)	(,) #___ (,)
(,) #___ (,)	(,) #___ (,)	(,) #___ (,)	(,) #___ (,)	(,) #___ (,)

<u>Notes</u>

_____

_____

_____

_____

Area:_____ Component: _____ Quantity: _____

(,) #___ (,)	(,) #___ (,)	(,) #___ (,)	(,) #___ (,)	(,) #___ (,)
(,) #___ (,)	(,) #___ (,)	(,) #___ (,)	(,) #___ (,)	(,) #___ (,)
(,) #___ (,)	(,) #___ (,)	(,) #___ (,)	(,) #___ (,)	(,) #___ (,)
(,) #___ (,)	(,) #___ (,)	(,) #___ (,)	(,) #___ (,)	(,) #___ (,)

<u>Notes</u>

_____

_____

_____

_____

SIGNATURE _____ DATE _____

GARDENS ON MARS!

PROJECT

Continued from page

Continued to page

SIGNATURE

DATE

Witnessed & understood by

DATE

Top Secret Information

GARDENS ON MARS!

PROJECT

Continued from page

Continued to page

SIGNATURE

DATE

Witnessed & understood by   DATE

Top Secret Information

Continued from page

Continued to page

SIGNATURE | DATE

Witnessed & understood by | DATE

GARDENS ON MARS!

Continued to page

SIGNATURE

DATE

Witnessed & understood by

DATE

GARDENS ON MARS!

Grid (x,y) Coordinates

Area:_____ **Component:** _____ **Quantity:** _____

(,) #___ (,)	(,) #___ (,)	(,) #___ (,)	(,) #___ (,)	(,) #___ (,)
(,) #___ (,)	(,) #___ (,)	(,) #___ (,)	(,) #___ (,)	(,) #___ (,)
(,) #___ (,)	(,) #___ (,)	(,) #___ (,)	(,) #___ (,)	(,) #___ (,)
(,) #___ (,)	(,) #___ (,)	(,) #___ (,)	(,) #___ (,)	(,) #___ (,)

<u>Notes</u>

_____

_____

_____

_____

Area:_____ **Component:** _____ **Quantity:** _____

(,) #_____	(,) #_____	(,) #_____	(,) #_____	(,) #_____	(,) #_____	(,) #_____
(,) #_____	(,) #_____	(,) #_____	(,) #_____	(,) #_____	(,) #_____	(,) #_____
(,) #_____	(,) #_____	(,) #_____	(,) #_____	(,) #_____	(,) #_____	(,) #_____
(,) #_____	(,) #_____	(,) #_____	(,) #_____	(,) #_____	(,) #_____	(,) #_____

<u>Notes</u>

_____

_____

_____

_____

SIGNATURE _____ **DATE** _____

GARDENS ON MARS!

PROJECT

Continued from page

Continued to page

SIGNATURE

DATE

Witnessed & understood by

DATE

Top Secret Information

Continued from page

Continued to page

SIGNATURE

DATE

Witnessed & understood by

DATE

Top Secret Information

GARDENS ON MARS!

PROJECT

Continued from page

Continued to page

SIGNATURE

DATE

Witnessed & understood by

DATE

Top Secret Information

GARDENS ON MARS!

Continued from page

Continued to page

SIGNATURE

DATE

Witnessed & understood by

DATE

Top Secret Informatio

Notes

_____

_____

_____

SIGNATURE _____ DATE _____

GARDENS ON MARS!

Continued to page

SIGNATURE

DATE

Witnessed & understood by

DATE

Continued from page

Continued to page

SIGNATURE

DATE

Witnessed & understood by

DATE

Top Secret Information

Continued from page

Continued to page

SIGNATURE

DATE

Witnessed & understood by

DATE

Top Secret Information

GARDENS ON MARS!

PROJECT

Continued from page

Continued to page

SIGNATURE

DATE

Witnessed & understood by

DATE

Top Secret Information

GARDENS ON MARS!
Grid (x,y) Coordinates

Area:_____ Component: _____ Quantity: _____

(,) #_____	(,) #_____	(,) #_____	(,) #_____	(,) #_____	(,) #_____	(,) #_____
(,) #_____	(,) #_____	(,) #_____	(,) #_____	(,) #_____	(,) #_____	(,) #_____
(,) #_____	(,) #_____	(,) #_____	(,) #_____	(,) #_____	(,) #_____	(,) #_____
(,) #_____	(,) #_____	(,) #_____	(,) #_____	(,) #_____	(,) #_____	(,) #_____

Notes

_____

_____

_____

_____

Area:_____ Component: _____ Quantity: _____

(,) #_____	(,) #_____	(,) #_____	(,) #_____	(,) #_____	(,) #_____	(,) #_____
(,) #_____	(,) #_____	(,) #_____	(,) #_____	(,) #_____	(,) #_____	(,) #_____
(,) #_____	(,) #_____	(,) #_____	(,) #_____	(,) #_____	(,) #_____	(,) #_____
(,) #_____	(,) #_____	(,) #_____	(,) #_____	(,) #_____	(,) #_____	(,) #_____

Notes

_____

_____

_____

_____

SIGNATURE _____ DATE _____

Notes

_____

_____

_____

SIGNATURE _____   **DATE** _____